The Carpenter's Codex

A Timeless Compendium of Traditional Carpentry Methods for Residential And Commercial Construction

Hazel Holcomb

Contents

1 Introduction to Carpentry **5**
1.1 Tools of the Trade 7
1.2 Safety First 10

2 Basic Joinery Techniques **13**
2.1 Mortise and Tenon 15
2.2 Dovetail Joints 17
2.3 Frame and Panel Construction 20

3 Framing Walls and Floors **25**
3.1 Platform Framing 27
3.2 Balloon Framing 30

4 Roof Construction **35**
4.1 Rafter Design 37
4.2 Truss Systems 39
4.3 Roof Sheathing 42
4.4 Finishing the Roof 44
4.5 Roofing Materials 46

5 Finishing Carpentry **49**
5.1 Installing Trim 51
5.2 Door and Window Installation 54

6 Advanced Carpentry Techniques 57
6.1 Timber Framing 59
6.2 Architectural Details 62
6.3 Restoration and Repair 65

7 Commercial Construction Carpentry 69
7.1 Large-Scale Framing 71
7.2 Specialized Techniques 74
7.3 Building Codes 76
7.4 Project Management 79
7.5 Safety Regulations 82

1 Introduction to Carpentry

Carpentry, at its core, is about shaping wood into functional and beautiful structures. It's a craft demanding precision, patience, and a deep understanding of the material. From the humble workbench to the soaring roofline of a cathedral, wood, when handled skillfully, transforms into works of art and functional spaces. The satisfaction derived from building something tangible, something that will last, is a unique reward. This chapter will introduce you to the basics, providing a solid foundation for your carpentry journey. This is where your passion will take shape and your skills will be sharpened.

Before you even pick up a saw, however, consider your workspace. A well-organized and well-lit area is crucial for safety and efficiency. Having your tools readily accessible minimizes wasted time and effort. Consider a dedicated workbench, preferably sturdy and large enough to accommodate your projects. Invest in good quality tool storage; keeping your tools organized not only aids efficiency but also extends their lifespan. Good lighting is essential; avoid shadows that can obscure your work and lead to mistakes. A poorly organized workspace leads to inefficiencies and accidents, therefore make organization your priority.

Safety is paramount in carpentry. This isn't just a suggestion; it's a necessity. Always wear appropriate personal protective equipment (PPE), including safety glasses, hearing protection, and work gloves. These aren't mere accessories; they are essential for protecting your eyes, ears, and hands from potential hazards. Remember, a single careless moment can lead to serious injury. Learn to use your tools correctly, following manufacturer instructions carefully. Never rush your work; haste increases the risk of mistakes and accidents. Proper safety practices aren't just about personal wellbeing; they create a more efficient workflow.

Understanding the basic tools is critical. Start with the essentials: a measuring tape, pencil, hand saw, hammer, chisels, and screwdrivers. These tools will form the foundation of your carpentry toolkit. Learn how to sharpen your tools properly, as sharpness significantly impacts both the quality and safety of your work. A sharp tool makes cleaner, more precise cuts and reduces the risk of accidents. Invest in high-quality tools; they will serve you well for years. Familiarize yourself with their use and maintenance; this will save you from many headaches down the line. Don't overload yourself; start with a few essential tools and gradually expand your collection as you gain experience.

Beyond hand tools, power tools significantly increase efficiency. A circular saw, for instance, enables quick and precise cutting of lumber, while a drill allows you to create pilot holes and fasten various materials effectively. However, remember to treat power tools with respect. Always read the instruction manuals thoroughly before use. Ensure the area is clear of obstructions, and always wear appropriate safety gear. Properly maintained power tools extend their lifespan and reduce the risk of accidents. Treat them with care to maximize their effectiveness and longevity; they are an investment that will pay off in the long run. Your journey into carpentry doesn't end here. This is just the beginning of a rewarding path. Learning and practicing are essential for mastery. Each cut, each joint, each completed project will contribute to your expanding skills. The process requires patience, perseverance, and a willingness to learn from your mistakes. As you progress, you will discover the sheer joy of creating with your own hands, transforming raw materials into something tangible, something beautiful, and something functional. Embrace the learning process, continually hone your skills, and enjoy the journey.

1.1 Tools of the Trade

A sharp chisel, its edge honed to perfection, is more than just a tool; it's an extension of your hand, your skill, your vision. Learning to properly sharpen and maintain it, understanding its nuances, becomes integral to the quality of your work. This isn't just about creating clean cuts; it's about respecting the tool, understanding its capabilities, and developing a feel for its responsive nature. A dull chisel, on the other hand, will frustrate you, lead to inaccurate cuts, and potentially damage the wood. It's a slow, inefficient, and potentially dangerous way to work. Treat your chisels with the respect they deserve, and they will reward you with precision and efficiency for years to come. Regular sharpening and oiling are crucial aspects of their maintenance.

Your measuring tape is your constant companion. It's more than a simple device; it's the cornerstone of accurate construction. More than just inches and feet, it represents precision, foresight, and planning. Never underestimate its importance—a mismeasurement can throw off an entire project. Learn to use it effectively; master the techniques of accurate measurement, avoiding common mistakes like measuring from the wrong point or failing to account for material thickness. Investing in a high-quality tape measure, one with a sturdy case and a clear, easy-to-read marking, is an investment in the long-term success of your projects.

The humble hand saw, with its teeth precisely set, is a tool that speaks to tradition and skill. It's not a relic; it's a precise instrument demanding finesse and understanding. Different saws are suited for different woods and cuts, each requiring a different technique. Practice makes perfect: learn to maintain a steady hand and consistent rhythm, ensuring clean and accurate cuts every time. Understanding the grain of the wood and adjusting your sawing technique accordingly will prevent splintering and ensure a smooth, professional finish. A sharp saw minimizes effort and maximizes precision, while a dull saw is inefficient, prone to error, and frankly, infuriating.

Beyond hand tools, power tools offer increased efficiency and precision when used correctly. The circular saw, for instance, demands respect and careful handling. Its speed and power require focus and adherence to safety guidelines. Learn its capabilities, limitations, and how to use various blades for different materials and cuts. Always use appropriate safety equipment, from eye protection to hearing protection, even dust masks, because the dust generated can be harmful. A poorly maintained saw or incorrect usage can result in inaccurate cuts and potentially severe injury.

A drill, whether corded or cordless, is essential for various tasks, from pilot holes to larger scale drilling. Understanding the different drill bits, their sizes, and their applications is critical. You'll need to choose the correct bit for the material and the job. Different materials require different speeds and pressures. Learn how to accurately locate holes and drill perfectly straight to avoid costly mistakes and potential damage to the material. Remember, practice and patience are key to mastering the effective use of power tools.

Safety should always be your foremost concern when wielding these tools of your trade. Always wear appropriate safety glasses or a face shield to protect your eyes from flying debris. Hearing protection is also essential, especially when using power tools. Understand the dangers associated with each tool and always follow manufacturer guidelines meticulously. This not only keeps you safe, but it also leads to more accurate and precise work, because you're not distracted by worries or discomfort. Safety isn't just a precaution; it's an essential aspect of a successful career as a carpenter.

Finally, remember that the tools are only as good as the person using them. Regular maintenance, understanding their limitations, and practicing good technique are vital. Investing in quality tools will pay off in the long run, both in the quality of your work and in your enjoyment of the craft. Your skill will only improve with time and practice, so don't be afraid to experiment, learn from your mistakes, and constantly strive to improve your technique. The path to mastery is a journey, not a destination.

1.2 Safety First

Working with wood, especially on construction sites, demands constant vigilance. Ignoring safety precautions isn't just reckless; it's a recipe for serious injury, even death. Remember, a single lapse in concentration can have devastating consequences. Always prioritize your well-being; it's the foundation upon which successful projects are built. The best projects are those completed safely. Never compromise your safety or that of your colleagues for the sake of expediency.

Eye protection is non-negotiable. Dust, flying debris, and even stray nails are ever-present dangers. Safety glasses should be worn at all times, regardless of the task at hand. Consider investing in a face shield for added protection during particularly hazardous operations like cutting or grinding. Protect your eyes; your vision is invaluable. Don't underestimate the importance of proper eye protection; a single piece of debris can cause irreversible damage.

Hearing protection is just as vital, particularly when operating power tools. Prolonged exposure to loud noises can lead to permanent hearing loss. Earmuffs or earplugs should be used consistently. Consider the cumulative effect of noise exposure; even seemingly brief periods of loud noise contribute to overall damage. Protecting your hearing now will safeguard your quality of life in the future. Invest in high-quality hearing protection; your hearing is too important to risk.

Respiratory protection is also crucial, especially when working with materials like sawdust or chemical treatments. A dust mask, or better yet, a respirator, can significantly reduce the inhalation of harmful particles. Consider the specific hazards associated with each project and choose appropriate respiratory protection accordingly. This protects your lungs and your overall health. Your health shouldn't be sacrificed for the job. Never underestimate the importance of clean air.

Appropriate clothing is more than just comfort; it's a critical safety element. Loose clothing can get caught in machinery, and open-toed shoes offer little protection from falling objects or sharp tools. Always wear sturdy closed-toe shoes and clothing that is well-fitting and durable. Avoid wearing jewelry or accessories that could present a hazard. Your attire is a vital component of your safety gear. Work clothes should be both protective and practical.

Handling tools correctly is fundamental to safety. Learn the proper techniques for using each tool, and always maintain your tools in good working condition. A sharp tool is a safer tool; dull tools require more force, increasing the risk of accidents. Regular maintenance is not an option, but an essential part of safe tool use. Keep your tools organized and easily accessible to prevent unnecessary reaching and fumbling.

Always be mindful of your surroundings. Know where your fellow workers are, and be aware of potential hazards, such as uneven ground or overhead obstacles. Maintain a clean and organized workspace; clutter increases the risk of accidents. A safe workspace significantly minimizes the chances of injury. Be vigilant and maintain awareness of your environment at all times. Constant observation is key to injury prevention. Before beginning any task, take a moment to assess potential risks. Identify hazards and determine the appropriate safety precautions. Plan your work carefully, and don't rush. Rushing leads to mistakes and injuries. Safe work practices are the backbone of accident prevention. Never begin any task until you understand all aspects of safety.

Finally, never hesitate to report any unsafe conditions. Your safety, and the safety of your coworkers, is paramount. Raising concerns is a sign of responsibility, not weakness. A safe work environment is a collaborative effort. Speak up – your safety matters. Remember that prevention is always better than cure. Your well-being is your most valuable asset.

2 Basic Joinery Techniques

The mortise and tenon joint, a cornerstone of woodworking, exemplifies strength and simplicity. Imagine two pieces of wood: one with a precisely cut hole, the mortise, and the other with a corresponding projection, the tenon. The tenon fits snugly into the mortise, creating a strong, almost inseparable connection. This joint, used for centuries in everything from furniture to structural beams, demonstrates the power of carefully planned geometry in construction. Its variations, from through tenons to wedged tenons, allow for adaptability to different stresses and aesthetic preferences. Mastering this joint is fundamental to any carpenter's skillset, as its reliability under load is unmatched by many other joinery methods. The precision required in its creation instills a respect for careful measurement and precise execution, lessons invaluable throughout your carpentry journey.

Dovetail joints, known for their elegant appearance and exceptional strength, demand both skill and patience. These interlocking finger-like projections, meticulously cut into mating pieces of wood, create a joint that resists both pulling and racking forces. The process itself is a testament to hand-craftsmanship, with each cut carefully planned and executed to ensure a perfect fit. Traditional hand tools are often preferred for creating dovetail joints, offering a level of control and precision difficult to replicate with power tools. This method instills a deeper understanding of the wood itself and the interaction between tool and material. While seemingly complex, mastering the dovetail joint not only improves woodworking technique but also enhances your creativity and design capabilities, allowing you to create beautiful and highly durable pieces. You'll find it applied in fine furniture, drawer construction, and other applications where both strength and visual appeal are paramount.

Frame and panel construction, a technique used extensively in furniture and doors, combines the strength of a sturdy frame with the lighter weight and often decorative quality of a central panel. The frame, typically made of stronger, often more visually appealing wood, is constructed using techniques like mortise and tenon or even dovetail joints. This creates a robust perimeter that supports the central panel. The panel itself might be constructed from less expensive or more easily worked materials. The panel is then carefully fitted into the frame, creating a strong and visually balanced piece. This technique allows for a combination of materials and aesthetic choices, making it extremely versatile. You can incorporate various woods, veneers, or even glass into the panel to achieve different aesthetic outcomes. Understanding frame and panel construction opens up a wider range of design possibilities and allows you to construct more complex pieces with confidence. It is a method that allows you to create attractive yet structurally sound pieces. Moreover, the ability to choose the frame and panel materials separately gives you greater control over the overall budget and aesthetic appeal.

2.1 Mortise and Tenon

The mortise and tenon joint, a cornerstone of traditional carpentry, represents a deceptively simple yet incredibly strong connection between two pieces of wood. Its enduring popularity stems from its robustness, ease of construction (once mastered), and aesthetic appeal. The fundamental principle involves a precisely sized hole, the mortise, cut into one piece of wood, and a corresponding projection, the tenon, crafted on the other. When these two parts are joined, they create an exceptionally strong and stable connection.

Think of it like a hand fitting perfectly into a glove. The precision of the fit is crucial. A loose fit compromises strength, while a too-tight fit makes assembly difficult and can lead to wood splitting. Proper tool selection and technique are therefore paramount in achieving a successful mortise and tenon joint. This includes using chisels, hand saws, and possibly a mortise machine for larger projects, each demanding precision and control.

Different variations exist, each offering unique properties and suited for specific applications. Through-tenons extend completely through the mortise, offering maximum strength ideal for heavy-duty applications like structural beams or table legs. This type of tenon provides optimal strength, resisting both shear and tensile forces. However, the through-tenon requires careful attention during cutting and assembly to prevent wood splitting.

In contrast, a blind mortise and tenon joint, where the tenon only partially enters the mortise, is frequently used in furniture making. The visible portion of the tenon is often left slightly proud, adding a nice visual accent and making the joint easily glued. Blind tenons are easier to execute for those still building their skills since the risk of wood splitting is reduced.

Choosing the right type of joint hinges on several factors. The wood's density, grain orientation, and the intended use of the piece all contribute to the final decision. Hardwoods tend to hold up better under stress, allowing for tighter fits in through-tenon applications. Softer woods, on the other hand, might require a slightly looser fit to prevent splitting during assembly. Consider also how much stress the finished piece will endure. For high-stress applications, through-tenons are superior while, for less demanding projects, blind tenons are more suitable.

Proper preparation is critical to success. Accurately marking out the mortise and tenon is essential. The slightest inaccuracy can compromise the strength and aesthetic appeal of the finished joint. Accurate marking typically involves using marking gauges, squares, and possibly a mortise gauge to ensure precise dimensions. Pre-drilling pilot holes before chiseling out the mortise reduces the risk of wood tear-out. Always work slowly and carefully and pay close attention to every step for a clean, effective joint.

The actual creation of the mortise and tenon themselves requires skill and patience. Chiseling a mortise involves careful, incremental removal of wood to achieve a precise, square hole. The tenon, meanwhile, is crafted with a fine saw and chisel to maintain precise dimensions and a clean, square shoulder. This step demands a sharp chisel and a steady hand for a cleanly cut mortise and a snug tenon. The goal is a precise fit where the joint should offer significant resistance to movement.

Once both mortise and tenon are prepared, the actual assembly is relatively straightforward. Wood glue is applied liberally to the tenon, and the assembly is tapped together until the joint is seated firmly. A clamp is then used to secure the joint, ensuring a proper setting and adequate drying time. Careful assembly, as well as a decent glue, are very important to the success of the mortise and tenon.

Finally, after glue has dried, the excess glue can be carefully removed. Any proud parts of the tenon can be trimmed or the entire joint can be carefully sanded smooth to blend seamlessly with the surrounding wood. This process refines the look of the joint and improves its final appearance. The final step, after the glue is fully cured, is to add a finish to protect the wood, enhancing both the durability and aesthetic appeal of the finished piece.

Mastering the mortise and tenon is a significant step in any carpenter's journey. It builds confidence, strengthens fundamental skills, and opens doors to more complex joinery techniques. With practice and patience, any aspiring carpenter can create strong, beautiful, and functional joints that stand the test of time. The investment in mastering this basic skill will pay dividends throughout your career, showcasing both craftsmanship and structural integrity. It is a foundation of traditional carpentry and an excellent skill to have, regardless of the size of your projects.

2.2 Dovetail Joints

The dovetail joint, a hallmark of fine woodworking, offers unparalleled strength and a visually stunning interlocking design. Its characteristic tapered pins and interlocking tails create a joint that resists racking forces far better than many alternatives, making it ideal for applications where strength and stability are paramount—think drawers, boxes, and even some furniture framing. This robust nature stems from the angled surfaces of the pins and tails, which interlock and grip tightly, increasing the joint's resistance to separation under stress. The more tails and pins used, the stronger and more secure the connection becomes.

Unlike simpler joints, the dovetail requires precision and careful planning. Accurate marking and cutting are essential to ensure a tight, seamless fit. Improperly cut dovetails will result in a loose or weak joint, potentially compromising the structural integrity of the piece. Therefore, sharp chisels and hand saws, or precise router bits, are crucial for achieving the necessary accuracy. The process involves laying out the tails on the workpiece and then marking and cutting the corresponding pins on the mating piece, which demands significant precision for a successful outcome.

Choosing the right type of dovetail is a crucial initial step. Through dovetails, where the pins and tails extend completely through the workpiece, are extremely strong but can be more challenging for beginners. Half-blind dovetails, with only the tails visible, offer a cleaner aesthetic, especially on the exterior of a piece. Then, there are full-blind dovetails, hiding both pins and tails for a completely seamless appearance. The choice depends on the specific project, the desired aesthetic, and the carpenter's skill level. Beginners often find half-blind dovetails a good starting point, offering a compromise between strength and ease of execution.

The process starts with careful marking. Layout tools, such as a marking gauge and a square, are invaluable in creating accurate and consistent measurements. Begin by marking the tail board's layout, ensuring the spacing between tails is even and the angle of the taper is consistent. The accuracy of this step directly affects the success of the entire joint. Once the tails are marked, carefully cut them, working slowly and methodically to avoid errors. Using a sharp chisel and applying gentle pressure is vital. Rushing this phase can lead to uneven cuts and a compromised joint.

After the tails are cut, transfer the layout to the pin board. This is best done using the tail board itself as a template, carefully pressing it against the pin board and marking the outline of each tail. Precise marking is crucial here; any inaccuracies will be directly reflected in the fitting of the pins. The pins, unlike the tails, are typically cut using a saw or router, carefully removing material to create the negative space for the tails. A sharp, finely-toothed saw is essential for achieving clean, accurate cuts. Test-fitting throughout the process is essential; any adjustments should be made before glue is applied.

Once the pins and tails are fully cut and fit, apply a high-quality wood glue. A thin, even layer is ideal. Carefully align the pins and tails, using clamps to hold the pieces firmly together until the glue has set completely. Clamping pressure should be even and consistent throughout the joint, preventing any gaps or misalignments. Removing excess glue is crucial to avoid ruining the project's appearance. After allowing the glue to fully cure, any excess glue can be carefully removed using a chisel or scraper.

Finally, after the glue is fully cured, carefully trim any excess wood protruding from the joint. This step requires precision to maintain the clean lines of the dovetail. Use a sharp chisel or plane to carefully level the joint, ensuring a perfectly flush surface. Proper finishing techniques, such as sanding and finishing with a suitable oil or varnish, enhance the appearance and protect the wood from the elements. The finished dovetail should be not only strong and structurally sound but also aesthetically pleasing, a testament to the carpenter's skill and precision. With practice, creating beautiful and durable dovetail joints becomes easier and more satisfying. The time and effort invested in mastering this fundamental joinery technique will pay dividends in numerous future projects.

2.3 Frame and Panel Construction

Frame and panel construction offers a robust and visually appealing method for creating doors, cabinet fronts, and other furniture components. The basic principle involves a sturdy framework, usually of thicker stock, which supports a thinner panel, creating a strong yet lightweight structure. Choosing the right wood for both frame and panel is crucial; hardwoods like oak or maple provide strength and durability for the frame, while softer woods like pine or poplar work well for panels, allowing for easier carving or shaping if desired. Proper wood selection significantly impacts the final product's longevity and aesthetic appeal. Accurate measurements are paramount. Imprecise cuts will lead to a wobbly or poorly fitting frame, undermining the entire structure's integrity. Use a quality measuring tape and double-check all measurements before making any cuts. This seemingly simple step is crucial, preventing wasted materials and frustrating rework. Remember, patience is key; rushing the process often leads to errors. Take your time and ensure accuracy in each step. Precision cutting, whether with hand tools or power tools, makes all the difference between a professional-looking finished product and a less impressive result.

The construction process begins with the frame. Typically, you'll create a rectangular frame using mortise and tenon joints, the most common method for durability and strength. Alternatively, you can use dadoes and rabbets for a simpler approach, but note that they're slightly less robust than tenons and mortises for high-stress applications. The choice depends on your desired level of strength and aesthetic appeal as well as the tools you have readily available. The frame components must be perfectly square and accurately sized. You'll often require precise angled cuts, so using a miter saw, miter box, or other specialized tools is crucial, especially when dealing with larger pieces of wood that may be difficult to manage with hand tools alone.

Once the frame is assembled, it's time to prepare the panel. The panel size should be slightly smaller than the internal dimensions of the frame, allowing for a snug fit. This panel can be solid wood, but plywood or medium-density fiberboard (MDF) are often preferred for their cost-effectiveness and stability, especially in larger panels. Regardless of the material choice, ensure the panel is planed smooth and free from any imperfections before installation, as these irregularities could compromise the final aesthetic of the project. A smooth panel makes a significant visual improvement to the finished piece.

Finally, the panel needs to be secured to the frame. This can be done in several ways, each with its own advantages. One common method is to use wood glue and clamps to secure the panel, ensuring an even pressure distribution across the entire surface. After gluing, carefully position the panel inside the frame, allowing the glue to form a strong bond. You'll likely need various clamps to apply sufficient even pressure. For additional strength and to conceal the attachment, small brads or finishing nails can be used, carefully countersinking them to avoid damaging the finished surface and allowing for proper filling and sanding later. Take care to avoid applying too much pressure, which could potentially cause the panel to warp or crack.

Finishing touches are essential. After the glue is fully dry, remove any excess glue and lightly sand any rough edges or imperfections in the wood to ensure an even surface for finishing. Wood filler can be used to fill nail holes or minor imperfections, ensuring a smooth surface for painting or staining. After sanding, apply a suitable finish—paint, varnish, stain—to protect the wood and enhance its appearance. This final step protects the frame and panel construction from moisture damage and enhances its overall aesthetics, ensuring a lasting and pleasing piece of woodworking. Remember, the right choice of finish greatly enhances the final product, so choose something that complements your design.

Properly constructed, a frame and panel assembly creates a durable and versatile component that can be incorporated into a wide variety of woodworking projects, from simple shelves to intricate doors or cabinets. The process may initially seem challenging, but with careful planning and attention to detail, even beginners can achieve professional-quality results. Mastering this technique provides a valuable skill in a carpenter's arsenal, opening up numerous creative possibilities. Consistent practice and attention to the fundamentals of measurement, joint construction, and finishing will ensure the highest quality results. The beauty and functionality of frame and panel construction is in its simplicity and strength. With practice and patience, you'll quickly master this fundamental technique and incorporate it into your own projects.

3 Framing Walls and Floors

Precise measurements are paramount. Begin by carefully laying out your floor joists, ensuring they're evenly spaced and perfectly aligned. Use a chalk line for accuracy; a slightly crooked floor will only compound issues later. Remember to account for any necessary adjustments for load-bearing walls or specific architectural details. Double-check your measurements before securing anything; correcting mistakes mid-project is far more difficult than preventing them in the first place. Using a level is essential, but trust your eyes too; sometimes, a slight imperfection is easier to spot visually than with a tool. Finally, remember to consider the type of subflooring you will be using; the spacing of your joists will influence the subfloor's stability and durability. The subfloor should be chosen to accommodate the final flooring material. Consider the building's design. Platform framing, with its horizontal layers, is simpler for beginners. Each level forms a solid platform before moving upward. This makes the process more manageable, especially for larger projects. Balloon framing, on the other hand, requires more experience and precision due to its continuous studs running from foundation to roof. Understanding how these methods differ is crucial. While balloon framing can save on material costs, its susceptibility to fire and its generally more complex nature makes it a less beginner-friendly approach. Always take time to plan out the structural support before cutting any lumber.

Wall framing follows a similar logic, with crucial considerations for structural integrity and load-bearing capacity. Remember to account for window and door openings. Cut studs precisely and ensure they are properly aligned and braced for optimal structural strength. Accurate measurements and cutting are crucial. Using a speed square will guarantee consistent and accurate cuts. Proper bracing during construction is imperative, especially for taller walls. Don't rush this step; taking your time here will prevent costly mistakes and safety hazards later. You'll need to consider the spacing of studs for insulation and wall sheathing as well. Using a framing square will help you measure and mark accurately, reducing potential errors. Always double-check your work; a minor inaccuracy early on can lead to significant problems down the line.

Accurate cutting techniques significantly impact the finished product. Using sharp tools is vital for clean cuts, preventing splintering and improving accuracy. Familiarize yourself with the capabilities of your saw, both hand saws and power saws. Learn to safely handle your tools. Improper techniques will lead to mistakes, potential injuries and wasted materials. Consider pre-cutting all of your studs to the same length whenever possible to ensure uniformity. Using jigs will improve accuracy and speed up the cutting process. Cutting the studs carefully to the correct length ensures efficient work and minimizes adjustments later on. This precision ensures a seamless and structurally sound frame. Finally, remember the importance of safety. Always wear appropriate safety gear, including safety glasses, hearing protection, and work gloves. Proper ventilation is also essential when working with power tools and wood dust. Never rush your work. Take breaks when needed to avoid fatigue. Use proper lifting techniques to prevent back injuries. Always maintain a clean workspace. A clean and organized area ensures safety and efficiency. Safety regulations can vary, so review the specifics for your region to avoid potential hazards. Remember, safety should be paramount, exceeding even the desire for efficiency. Taking safety shortcuts can lead to serious accidents.

3.1 Platform Framing

Platform framing, the dominant method in modern residential construction, relies on a layered approach. Each floor acts as a platform for the next, creating a robust and stable structure. This method offers significant advantages in terms of speed and efficiency, making it the preferred choice for many builders. Understanding its principles is crucial for anyone hoping to excel in carpentry, whether for personal projects or professional work. The system's simplicity allows for straightforward construction, even for relatively inexperienced carpenters, while allowing for easy modification and expansion should the need arise. This adaptability makes platform framing an excellent foundation for diverse building projects.

Unlike balloon framing, which utilizes continuous studs extending from foundation to roof, platform framing creates distinct levels. This segmented approach offers better lateral stability, crucial for withstanding seismic activity or strong winds. It also provides fire resistance, as the fire is contained within one section. The horizontal framing of each floor provides a sturdy platform, simplifying the process of erecting walls. This layered approach creates a more predictable and manageable construction process, leading to fewer errors and faster completion times. Proper planning and accurate measurements are still paramount. Even though this style offers many advantages, a lack of planning and attention to detail can lead to issues later on.

Accurate cutting is paramount in platform framing. The use of a level is crucial to ensure that each layer is perfectly aligned and square. Even slight discrepancies can accumulate throughout the structure, leading to significant problems later on. Precise measurements are not merely best practice, but a fundamental necessity. It's the difference between a sturdy, well-built home and one that's prone to shifting or settling. We've all seen those houses with slightly crooked walls \item that's a result of a lack of precision during the platform framing stage. Therefore, take your time and double-check your measurements before making any cuts.

The construction of walls on a platform frame begins with the floor joists, which form the base. These joists are typically spaced at 16 inches or 24 inches on center, depending on local building codes and the intended load-bearing capacity. Properly attaching these joists to the foundation is equally important to the structural integrity of the entire building. Next, the subflooring is laid, creating a solid and level surface for the walls. Subflooring provides a strong surface for walking, and is also extremely helpful in keeping everything aligned and in place. While seemingly straightforward, paying attention to proper nailing patterns and consistent spacing is key to preventing squeaking and movement within the floor. This seemingly small detail can impact the finished building more than people may think.

Wall construction involves creating a framework of vertical studs, typically 2x4s or 2x6s, depending on building codes and local conditions. These studs are spaced at 16 inches or 24 inches on center, matching the joist spacing. Top plates are then added horizontally, completing the wall framework. Sheathing, typically plywood or oriented strand board (OSB), is attached to the exterior of the walls, providing structural rigidity and a surface for the exterior cladding. It's during this stage that things like plumbing and electrical work are roughed in, working directly with the framing elements. Ensuring correct placement and protection of these is also part of the platform framing process. Once the sheathing is applied, and the walls are secured, there's a greater sense of structure and it's a rewarding experience.

The process is repeated for each floor level, with each floor acting as the platform for the next. This layering provides stability and makes the construction process easier to manage. Each stage builds upon the previous, creating a cumulative effect. Therefore, care in one stage directly impacts the ease and quality of the next. This cascading effect emphasizes the importance of precise work from the beginning until the final stages of the building. If there is a mistake during any of the layers, it can affect the whole building. You need to make sure that every detail and part is properly placed.

Proper nailing is critical throughout the entire platform framing process. Using the correct size and type of nails, driven straight and deep, ensures strength and prevents future movement. It's a simple element, but it's one that's easy to overlook. Always follow manufacturers' recommendations and building codes regarding nail spacing and penetration depth. Ignoring these seemingly minor details can have significant consequences later on, potentially compromising the structural integrity of the building. Therefore, pay close attention to detail at each stage.

Finally, understanding local building codes and regulations is essential. These codes vary from location to location, dictating things like lumber size, spacing, and nailing requirements. This is something often overlooked. Failing to adhere to local codes can result in legal issues and possibly even structural failure. Always check with your local building department before starting any project to ensure your work complies with all relevant regulations. It's one thing to build a solid structure, but it is another thing to make sure that it respects all the laws and codes. This is important because, after all, the finished project needs to last for a long time, and should not present any danger to the occupants of the building.

3.2 Balloon Framing

Balloon framing, a method predating platform framing, presents a distinct approach to wall construction. Instead of building walls in sections, studs extend continuously from the foundation sill to the roof plate, creating a unified structure. This uninterrupted verticality offers some advantages, particularly in terms of strength and resistance to racking, a sideways force that can destabilize a structure. Imagine a single, tall tree, its trunk representing the continuous studs, far more resistant to wind than a collection of shorter, independently supported branches. However, this inherent strength comes with its own set of challenges and considerations.

The continuous nature of the studs in balloon framing facilitates a more efficient use of materials, potentially reducing the overall lumber requirement compared to platform framing. Fewer plates and sills are needed, translating to material cost savings and decreased labor. Furthermore, this method can be beneficial in taller buildings, providing inherent strength to counter gravitational forces and wind load. Think of it like a tall ship's mast, held firmly by the entire structure, rather than merely resting on a separate platform. The reduced number of horizontal members also simplifies the framing process, leading to faster construction in some scenarios.

However, the uninterrupted length of the studs presents specific difficulties. Firstly, the long continuous studs are more susceptible to fire damage, creating a path for flames to travel easily from floor to floor. This necessitates extra attention to fire-stopping and fire-retardant treatments. It's a critical safety consideration that demands careful planning and meticulous execution. Secondly, the taller, continuous studs make the process of framing significantly more demanding. Precise measuring and cutting are paramount to ensure stability. Any errors in cutting or placement are magnified significantly, potentially affecting the overall stability of the structure. Maintaining plumb and square becomes considerably more challenging.

Furthermore, balloon framing presents challenges during the construction process. The longer studs are more difficult to handle and install, requiring more expertise and care to avoid damage. Access to the wall cavities is limited once the building is enclosed, making it more complex to run plumbing or electrical wires. This increases labor costs and the complexity of the overall construction. You need to pre-plan meticulously, carefully routing and installing utilities within the confined spaces before the building is completely framed. This intricate process necessitates careful planning and execution, demanding a higher level of skill and experience than platform framing often does.

Despite these challenges, balloon framing remains a relevant technique, especially in situations where the benefits outweigh the complications. It shines in situations requiring exceptional height, where minimizing horizontal members is beneficial, and where the increased strength is crucial. For instance, you might find it advantageous in barns where increased vertical strength is desired to support heavy loads, or in some multi-story residential buildings, depending on regional codes. Understanding both the advantages and disadvantages, however, is crucial for making an informed decision regarding its suitability for a specific project.

Experienced carpenters understand that material choices heavily influence the effectiveness of balloon framing. Using high-quality, appropriately treated lumber is paramount to mitigating fire risks and ensuring the structural integrity of the building. Additionally, meticulous attention to detail is crucial. Properly fitting fire stops at every floor level is not just a good practice; it's a crucial safety measure preventing a catastrophic fire from easily spreading throughout the entire building. Every aspect of the construction, from the initial layout to the final inspection, needs dedicated attention to minimize potential problems and maximize the inherent strengths of this older framing method. Proper bracing and shear wall placement are paramount to guarantee the overall stability.

Ultimately, the decision of whether to use balloon or platform framing hinges on a number of variables: the building's height, intended use, local building codes, and the carpenter's experience. While platform framing may be more common today, due to the ease of construction and improved fire safety, balloon framing retains its niche. Its advantages in certain situations are undeniable, and mastering this technique enhances a carpenter's skillset significantly. It's a testament to the enduring value of understanding traditional construction methods alongside modern techniques. Careful consideration of all the factors, both advantages and disadvantages, will always guide a carpenter to the best solution for a particular project, building on established traditional methods.

4 Roof Construction

Careful planning is the cornerstone of successful roof construction. Begin by thoroughly reviewing the architectural plans, noting roof pitch, overall dimensions, and the location of any penetrations like chimneys or skylights. Accurate measurements are paramount; even a slight error can lead to significant problems later on. Double-checking your calculations and using a variety of measuring tools can help avoid these preventable mistakes. Remember, precision is not just about accuracy, it's about building a safe and structurally sound roof.

Understanding rafter design is crucial. You'll need to calculate the length of each rafter considering the roof pitch and overhang. Many online calculators and software programs can help with this, but understanding the underlying principles—especially for more complex roof designs—is vital. Remember to factor in the additional length needed for overhangs and the type of rafter tail you'll be using. Don't skip this crucial step; getting the rafter length wrong can severely compromise the entire roof structure. Consider the type of wood you'll be using; pressure-treated lumber might be more resistant to rot.

Truss systems offer a prefabricated alternative to traditional rafter construction. These engineered components provide a strong and efficient solution, particularly for larger spans or complex roof designs. However, proper installation is vital; misaligned trusses can lead to structural instability. Ensure proper bracing and support throughout the installation process, particularly when working at height. Always follow the manufacturer's instructions precisely, and don't hesitate to consult with a structural engineer if you encounter unexpected challenges. Remember that safety is your number one priority.

Once the rafters or trusses are in place, sheathing forms the crucial weather barrier. Plywood or OSB panels are commonly used, requiring proper alignment and fastening to ensure a strong and watertight surface. Pay close attention to the spacing of fasteners; under-fastening can lead to sagging, while over-fastening can split the sheathing. Ensure proper ventilation by leaving small gaps for air circulation to prevent moisture buildup. A well-sheathed roof is fundamental to the longevity and structural integrity of your project.

Finishing the roof involves multiple layers of protection. The underlayment acts as a secondary weather barrier, protecting against wind-driven rain and snow. It should be carefully overlapped and sealed to prevent leaks. The choice of roofing material – asphalt shingles, tiles, metal, or other options – will influence both cost and aesthetics, and should suit your building style and climate. Remember the proper installation procedures for whichever roofing material you select; if incorrectly installed, even high-quality roofing materials will fail. Also, use quality materials for flashings around chimneys and other roof penetrations to avoid leaks.

Proper ventilation is crucial for a long-lasting roof. It helps regulate moisture levels, preventing condensation that can lead to rot and structural damage. Install ridge vents, soffit vents, and possibly gable vents to create a balanced airflow. Avoid overcrowding your roof space with insulation; it can impede the flow of air, leading to the very problems you are trying to prevent. Always consult building codes when installing ventilation systems; following these regulations will ensure optimal ventilation within the structure. A well-ventilated roof can significantly extend the life of your roof and protect the overall integrity of the building.

4.1 Rafter Design

Determining rafter length is fundamental. You'll need the run (horizontal distance from the wall to the ridge), and the rise (vertical distance from the wall plate to the ridge). Use the Pythagorean theorem—$a^2 + b^2 = c^2$—where 'a' is the rise, 'b' is the run, and 'c' is the rafter length. Remember to account for overhang; add this to your calculated rafter length. Accurate measurement is critical; even small errors can significantly impact the final structure's integrity. Always double-check your calculations and measurements before cutting.

Understanding rafter angles is equally important. The angle is determined by the rise and run, influencing the pitch of your roof. Steeper pitches shed snow better but require more materials. Gently sloped roofs use less material but might not perform as well in heavy snow. Consider local climate and building codes when choosing your roof pitch; these often dictate minimum and maximum slopes for structural stability and weather resistance.

Cutting rafters accurately requires precision and the right tools. A sharp hand saw or a power miter saw is crucial for achieving clean, precise cuts. Use a speed square or framing square to mark your cuts accurately according to your calculated angles. Paying attention to detail here will save you significant time and effort later on, and prevent costly mistakes. Remember to always double-check your measurements before making any cuts.

Proper rafter placement is crucial for a strong and stable roof. Spacing is typically determined by local building codes and the size of your roof sheathing. Consistent spacing ensures even load distribution, vital for preventing structural issues. Utilize a plumb bob to ensure rafters are perfectly vertical and aligned. This will ensure the roof is plumb and level, preventing future problems.

Connecting rafters to the wall plates and ridge board requires strong and reliable joinery. Traditional methods involve using a bird's mouth cut at the wall plate and a ridge cut at the peak. These cuts create a strong connection, transferring loads effectively. Consider using metal connectors for additional strength, particularly in areas with heavy snow or wind loads. These connectors offer added security and are essential in certain climates. Always follow local building codes regarding required fasteners and connections.

Remember to consider additional structural elements. Collar ties, for example, help prevent rafter spread and increase overall roof stability. They connect opposing rafters, preventing outward pressure. Consider adding purlins for larger spans. These horizontal members add further support and distribute the load more evenly. Proper use of these elements is paramount in ensuring your roof's structural soundness. Always factor in any additional weight when considering the overall

strength of your construction.
Don't overlook the importance of proper ventilation. Adequate ventilation helps prevent moisture buildup, which can lead to rot and damage. Consider installing vents at the eaves and ridge to allow for airflow. Proper ventilation is crucial for prolonging the life of your roof, saving you money and time on future repairs. This aspect is often overlooked but it is integral for long-term structural integrity.
Finally, work safely. Always use appropriate safety equipment such as safety glasses, gloves, and hearing protection when working with power tools. Be aware of your surroundings and ensure a stable work environment. Take your time and focus on accuracy; rushing can lead to errors and accidents. Safety should always be your top priority, preventing injuries and ensuring a safe working environment. Remember, careful planning and methodical execution are crucial for the successful construction of your roof.

4.2 Truss Systems

Trusses are prefabricated assemblies of lumber, designed to efficiently support roof loads. Unlike traditional rafter systems which require extensive on-site cutting and fitting, trusses are manufactured off-site to precise specifications, offering significant time and cost savings. This precision also enhances structural integrity, resulting in a stronger and more consistent roof structure overall. This makes them a popular choice for both residential and commercial construction projects, particularly those needing rapid construction time or where complex roof designs are involved. Understanding their design and construction is crucial for any carpenter involved in roof framing.

The basic components of a truss include chords (top and bottom members), webs (vertical and diagonal members), and connecting plates. The top chord usually bears the majority of the tensile load. The bottom chord absorbs the compression loads and is a critical element in maintaining the truss's overall strength. Web members provide crucial support, transferring the load from the top chord to the bottom chord and stabilizing the entire structure. These members are carefully engineered to withstand the forces they will face; their precise angles and placement are critical. Incorrect angles or member lengths will compromise the entire structure, often with disastrous consequences.

Metal connectors, commonly referred to as gusset plates, are essential components in truss construction. These plates, typically made of galvanized steel, are used to connect the various members of the truss system creating a rigid assembly. The strength and durability of the truss depend heavily on the quality and proper installation of these plates. Ensure that you use plates specifically designed for truss construction, and always follow the manufacturer's instructions for installation. Improper installation can lead to structural weakness. Furthermore, always use the correct size and type of fasteners – using undersized or incorrect fasteners will weaken the connection.

Different truss types cater to various roof styles and spans. Common types include gable trusses, hip trusses, and scissor trusses; each with unique strengths and structural characteristics. Gable trusses, characterized by their triangular shape, are suited for simple gable roofs and offer straightforward load distribution. Hip trusses, with their sloping sides on all four sides, are commonly found on more complex roof structures. Scissor trusses, featuring an inverted V shape, are often employed for higher ceilings and offer a visually appealing interior aesthetic, although they might demand a more skillful installation. Choosing the appropriate truss type depends entirely on the specific requirements of the project.

Proper handling and installation of trusses are crucial to avoid damage. Trusses are relatively lightweight but can still be bulky and cumbersome to handle, making correct lifting and placement important. Always use the appropriate lifting equipment, such as a crane or forklift, to avoid bending or damaging the members. Follow the manufacturer's instructions meticulously during placement and fastening. Never attempt to modify a truss after it is manufactured unless you have a thorough understanding of structural engineering principles and the implications for the system's strength. Improper modifications can severely compromise the structural integrity.

Understanding the engineering behind trusses and how loads are distributed is key. While this often involves complex calculations, a basic grasp of load paths is essential for safe and effective truss installation. Loads are transferred from the roof covering (shingles, tiles, etc.), through the sheathing and rafters to the supporting walls. The trusses themselves support the roof load and this is then spread to the exterior walls and finally down to the foundations. It's essential to ensure a solid foundation to support the entire weight effectively and ensure long term stability. Understanding this load path helps ensure the entire structure remains stable and safe.

Finally, remember the importance of local building codes and regulations. These codes dictate the acceptable truss designs and installation methods for your area, considering local factors like snow load and wind speed. Always check with your local authority for specific requirements before starting any truss installation. Ignoring these requirements could lead to rejection of your work, significant rework, and legal issues. Compliance with building codes ensures structural integrity and protects both the builder and the occupants of the structure from any potential problems. Remember, safety and adherence to regulations always come first when working with structural elements like trusses.

4.3 Roof Sheathing

Plywood sheathing forms the crucial underlayer for your roofing system, providing a stable and even surface for the final roofing material. Choosing the right plywood is paramount; it needs to withstand the stresses of the roof's weight and weather conditions. Consider the span between your rafters; longer spans demand thicker, stronger plywood to prevent sagging. Don't skimp here – the strength of your roof directly relies on a properly chosen and installed sheathing. Remember that the grade of plywood impacts its durability and longevity; opt for a higher grade for greater resistance to moisture and damage.
Proper fastening is just as important as selecting the correct plywood. Using the right nails or screws, driven at the correct depth and spacing, ensures a secure bond. Poorly fastened sheathing will lead to creaks, squeaks, and potentially structural weakness. Follow manufacturer recommendations diligently regarding fastener type, spacing, and penetration depth. Pre-drilling pilot holes for screws in denser plywood can prevent splitting and maintain the integrity of the sheet. Never under-

estimate the importance of thorough and accurate fastening.

Before you begin, accurately measure and cut your plywood sheets to fit the roof's dimensions. Precise cuts prevent gaps and overlaps, maximizing efficiency and minimizing waste. Utilize a circular saw or a hand saw for clean, accurate cuts; a skill saw makes faster work of larger sheets, but necessitates caution to maintain safety. Remember to consider the direction of the wood grain and the orientation of the sheathing sheets for optimal structural performance and to minimize potential warping. Accurate cuts are crucial for a seamless, structurally sound roof sheathing.

Laying the sheathing requires planning and attention to detail. Start at the bottom edge of the roof, ensuring the first sheet is perfectly aligned and securely fastened. Maintain consistent overlapping between sheets, typically 4 to 6 inches, creating a continuous, sturdy layer. This overlap ensures proper coverage and prevents water penetration. Always check for any inconsistencies, adjusting as needed, and ensuring that the roof structure can adequately support the added weight. This meticulous process avoids creating weak points in the roof's structure. A carefully laid sheathing is the bedrock of a strong and long-lasting roof.

Once the sheathing is complete, thoroughly inspect your work for any loose fasteners, gaps, or damaged sections. Addressing any problems early prevents more extensive repairs later. Examine the alignment of each sheet to ensure consistent overlapping and coverage. Repair or replace any damaged or inadequately secured sections. A final thorough check ensures a reliable and long-lasting roof structure. Don't hesitate to revisit and double-check your work; preventative measures save time and money in the long run. A sound roof sheathing ensures the integrity of your entire roofing system, directly impacting the longevity and structural soundness of the building.

4.4 Finishing the Roof

The final stage of roof construction demands meticulous attention to detail. Proper flashing installation is paramount; this thin, flexible metal prevents water intrusion at critical junctions like chimneys, vents, and valleys. Neglecting flashing invites leaks, leading to costly repairs down the line. Remember to overlap flashing pieces generously, ensuring a watertight seal, and use appropriate sealant where necessary for added protection. Don't skimp on this crucial step; a well-installed flashing system is your roof's first line of defense against the elements. Think of it as the roof's insurance policy.

Counterflashing, a secondary layer of flashing installed beneath the primary flashing, provides an extra layer of security, especially around chimneys. This dual defense system effectively diverts water away from vulnerable areas, significantly reducing the risk of leaks. This is particularly important in areas with heavy rainfall or snowfall, where water pressure can test the limits of even the most robust roofing systems. Always prioritize thoroughness; a small oversight in flashing can lead to significant problems.

Next comes the installation of the underlayment. This waterproof membrane acts as a secondary barrier against moisture, protecting the structure even if the roofing material fails. Properly overlapping the underlayment sheets is crucial, ensuring a continuous, seamless layer. Remember to secure it with appropriate fasteners, ensuring it remains taut and firmly attached to the roof deck. Don't allow any gaps or loose sections, as these could easily become entry points for water.

Now you are ready for the final layer: the roofing material itself. The selection of roofing material depends on numerous factors, including budget, climate, and aesthetic preferences. Asphalt shingles remain a popular choice for their affordability and ease of installation. However, more durable options such as tile, metal, or slate offer greater longevity and resistance to harsh weather conditions. Careful consideration should be given to the local climate and building codes when making this crucial selection. It's an investment that pays for itself over the lifespan of the building.

Once the roofing material is installed, carefully inspect the entire roof for any imperfections or areas that need attention. Minor adjustments or repairs should be made promptly to avoid future problems. A thorough inspection ensures your hard work results in a leak-free and aesthetically pleasing roof. This meticulous step not only safeguards the building but also protects your reputation as a skilled craftsman. You want a result that reflects your skill and pride in your work.

Finally, carefully clean the job site. Removing debris, discarded materials, and leftover tools contributes to a professional finish. This small, but significant action, showcases a commitment to quality that extends beyond the final nail. It is as much a testament to the pride of your work as the roof itself. Remember, the final touches often leave the most lasting impression. The quality of the cleanup will always resonate in the minds of the homeowner. The final impression reflects your quality and professionalism.

4.5 Roofing Materials

Asphalt shingles are the most common roofing material for residential construction due to their affordability and ease of installation. They're available in a wide range of colors and styles, offering versatility for various architectural designs. However, their lifespan is generally shorter than other options, typically lasting 15-30 years depending on quality and climate. Regular maintenance, such as cleaning debris and inspecting for damage after severe weather, can extend their useful life. Remember, proper ventilation beneath the shingles is crucial to prevent moisture buildup and premature deterioration.

Wood shingles and shakes offer a classic, aesthetically pleasing look, particularly suited for homes with traditional or rustic styles. These natural materials provide excellent insulation and can last for decades with proper maintenance, but they are more expensive and require more care than asphalt shingles. They are susceptible to rot, insect infestation, and fire, necessitating regular treatments with fire retardants and preservatives. Furthermore, the installation process is more complex and requires skilled labor, which increases the overall cost. Consider their higher initial cost and maintenance requirements before making a choice.

Clay tiles, with their durability and longevity, represent a significant investment. They offer exceptional weather resistance, fire resistance, and a long lifespan, often exceeding 50 years. Their weight necessitates a strong roof structure capable of supporting their considerable mass. Aesthetically, they're highly attractive, providing a timeless and elegant look. However, they're significantly more expensive than asphalt or wood shingles, and their installation demands expertise, driving up the labor costs. Their fragility during installation also demands extra care.

Metal roofing, encompassing materials like aluminum, steel, and copper, has risen in popularity due to its long lifespan, superior durability, and energy efficiency. Metal roofs are highly resistant to extreme weather conditions, including strong winds, heavy snow, and hail. They are also energy-efficient, reflecting solar radiation and reducing cooling costs. Their longevity can reach 50 years or more, significantly surpassing other roofing materials. However, the initial cost is generally higher, and they can be noisy during rainfall. The aesthetic appeal is highly dependent on the type and color chosen, ranging from sleek and modern to rustic and traditional.

Slate roofing, while undeniably beautiful and exceptionally long-lasting, often exceeding 100 years, comes with a substantial price tag. Its weight demands a robust roof structure and skilled installation, making it a high-cost option suitable for larger and more substantial buildings. Slate offers unmatched durability, resisting extreme weather and aging gracefully over time. The aesthetic appeal is undeniable, with its sophisticated and timeless beauty. However, its fragility during transport and installation necessitates specialized handling and skilled installers. Consider this option for its unparalleled lifespan and aesthetic value only after factoring in the substantial cost considerations.

Choosing the right roofing material involves careful consideration of several factors. Budget is a major constraint, with some options far more expensive than others. Your home's architectural style should dictate aesthetic choices; certain materials suit certain styles better. The local climate, with its typical weather conditions, greatly impacts material choice. A region with heavy snowfall requires a roof capable of withstanding substantial weight. Similarly, areas prone to extreme winds need materials that can resist them. Finally, the anticipated lifespan plays a vital role; some materials offer significantly longer service life than others. All of these aspects should be weighed carefully before making a final decision.

5 Finishing Carpentry

The final touches are what transform a house into a home, a building into a masterpiece. Finishing carpentry isn't just about aesthetics; it's about precision, attention to detail, and the satisfaction of seeing your hard work culminate in a flawless finish. It's the difference between a functional structure and a truly impressive one. We've laid the foundation, framed the walls, and built the roof; now, we'll add the finishing flourishes.

Installing trim is more than just nailing molding to a wall; it's about understanding the subtle nuances of angles, miters, and coping. Consider the type of wood: hardwoods like oak or cherry offer durability and elegance, while softer woods like pine are more affordable but require careful handling to avoid dents and scratches. The profile of the trim itself is crucial; simple crown molding offers a clean line, while more ornate designs add visual interest. Proper measurement is paramount; inaccuracies will be glaringly obvious once installed. Pay close attention to the alignment of each piece, ensuring tight joints and a seamless flow. Finally, the finishing touches—caulking, sanding, and painting or staining—transform the raw wood into a polished accent.

Accurate door and window installation is critical for both aesthetics and functionality. A poorly installed door will never close properly, leading to drafts, energy loss, and potential security vulnerabilities. It begins with careful preparation. Check the frame for plumb and square, ensuring a level surface for perfect alignment. Accurate measurements are essential to prevent gaps or binding. Use shims to ensure the door or window is perfectly plumb and square within the frame, ensuring smooth operation and a tight seal. This often involves fine adjustments, patient checking, and the satisfying feeling of perfect fit. Then, secure the unit firmly, using appropriate fasteners. Finally, install weather stripping and check for air leaks. The quality of the installation will directly impact the energy efficiency and longevity of your project.

Beyond doors and windows, consider baseboards, casings, and chair rails. Each piece presents its own set of challenges and rewards. Baseboards protect walls from damage and add a finishing touch to the room. Careful measurements, precise cuts, and neat caulking create a professional and polished look. Similarly, casings around doors and windows provide a clean, finished look and protect the surrounding walls. Consider the style of your home and choose accordingly; simple casings work well in modern homes, while more ornate styles suit traditional architecture. Chair rails add a touch of sophistication, protecting walls from chair backs while providing a decorative element.

Remember to always use the correct tools and materials. The right saw for the job will make a world of difference in the precision and quality of your cuts. Use appropriate fasteners, ensuring they are long enough to penetrate the wood securely, but not so long that they poke through. Always use a level and plumb bob to ensure accuracy. And don't forget safety; always wear safety glasses, hearing protection, and a dust mask

when necessary. These are not just recommendations; they are vital for preventing injuries and producing high-quality work.
Finishing carpentry is where your project truly comes alive. It's a testament to your skill and dedication. It's the culmination of countless hours of hard work, precision cuts, and careful planning. Taking the time to perfect these details will significantly enhance the overall appearance and value of your finished project. More importantly, it's the reward of creating something beautiful and enduring. And the pride you take in your craftsmanship is its own kind of reward, something far beyond the monetary value. This careful attention to detail transforms a simple structure into a home filled with pride, comfort, and lasting quality, a testament to your skills. Each precise cut, each carefully installed piece, reflects your dedication to excellence. The finished product isn't merely a house or building; it's a testament to your dedication and an enduring expression of your artistry.

5.1 Installing Trim

Installing trim is the final flourish, the jewelry that elevates a house from a structure to a home. It's about more than just nailing boards; it's about precision, attention to detail, and understanding the subtle nuances that make all the difference. Before you even think about grabbing your nail gun, carefully plan your approach. Measure twice, cut once—this age-old adage is especially crucial here. Accurate measurements prevent gaps and ensure a seamless, professional look. Consider the type of trim – crown molding, baseboards, casings – and their specific requirements. Each type presents unique challenges and necessitates a slightly different approach.

Crown molding, for instance, often involves intricate miter cuts at corners. Mastering these cuts requires practice, a sharp saw, and a miter box or miter saw—all tools you should be well-versed in using from previous chapters. Don't underestimate the importance of a perfectly aligned miter cut; a slight error here can ruin the entire installation. Remember that precision is paramount; slight inconsistencies will be noticeable. Use a high-quality wood glue along with your fasteners for extra strength and to prevent movement over time. For large projects, you might consider using a nail gun, but for more intricate work, hand nailing might be necessary for better control.

Baseboards, while seemingly simpler, present their own set of challenges. They often require precise cuts around doorways and windows, accounting for uneven walls and floors, which you've hopefully already addressed prior to this stage. Using shims to fill gaps between the baseboard and the wall is a common practice. Shims provide the necessary adjustment to ensure a snug, gap-free fit, avoiding any ugly spaces where dust and debris can accumulate. However, ensure the shims are securely fixed to avoid future issues and potential damage to the trim. For a truly professional finish, caulk the gaps between the baseboard and the wall. This not only seals the gap but also prevents drafts and improves the overall aesthetic appeal of your work.

Window and door casings add that polished touch. These casings should be meticulously measured and cut to create a smooth transition between the window or door and the surrounding wall. The accuracy and neatness of these cuts are directly proportional to the overall impact of your efforts. Remember to consider the type of wood used for the casings, ensuring that the wood is appropriately treated to resist moisture and maintain its integrity for years to come. Proper installation involves precisely positioning the casings and securing them using appropriate nails or screws—avoiding any damage to the underlying surface.

Beyond the actual installation, proper finishing is critical. After installing the trim, take time to fill nail holes and sand any rough edges smooth. This creates a uniform, professional surface for painting or staining. Once sanded, your trim will be ready for the next stage—applying primer and your chosen finish. A consistent application of finish, be it paint or stain, is critical in achieving a uniform and professional result. Make sure the finish you choose is compatible with the type of wood you're working with to avoid unexpected problems.

Remember, installing trim is a journey of precision and patience. It requires careful planning, meticulous measurements, and attention to detail. While it's the final step in many carpentry projects, it's a crucial one, adding the finishing touch that transforms a functional space into a beautiful home. Don't rush the process. Take your time, be precise, and pay attention to the small details. Your effort will be rewarded with a result that elevates your work to a whole new level. Take pride in your work and always use the best quality materials for a lasting result. The effort invested in this final stage is well worth it, reflecting your commitment to craftsmanship and creating an enduring result that stands as a testament to your skill. This careful attention to detail transforms a mere house into a true home, a space that reflects not just skilled craftsmanship but also dedication and pride in one's work. This is the essence of carpentry.

5.2 Door and Window Installation

Precise door and window installation is crucial for both energy efficiency and the overall aesthetic appeal of any building. Begin by carefully checking the rough opening; it must accurately match the manufacturer's specifications for your chosen door or window unit. A poorly sized opening will lead to gaps, drafts, and compromised security, ultimately impacting the structural integrity of the building. Using a level to ensure the opening is perfectly plumb and square is paramount before you even think about installing the unit. Inaccurate measurements at this stage will result in a difficult, time-consuming, and potentially costly fix later.

Next, consider the type of framing you're working with. Platform framing typically requires a different approach compared to balloon framing. In platform framing, you'll need to ensure the header is properly supported and adequately sized to bear the weight of the structure above the opening. Accurate header placement is important for a snug fit and prevents future settling issues; a poorly placed header can lead to structural instability. In balloon framing, where studs continue from the foundation to the roof, the process may involve careful consideration of how the new framing interacts with the existing long studs. Understanding these nuances and adapting your techniques accordingly is essential.

Once the opening is prepared, you can then prepare the window or door unit for installation. Carefully review the manufacturer's instructions to ensure you understand the unique installation requirements of your specific model. This may include specific shimming techniques, sealant applications, or special hardware considerations that are crucial for the unit's optimal performance and longevity. Ignoring these instructions could void warranties or lead to operational problems.

Installation itself involves carefully positioning the unit within the rough opening. Accurate placement is crucial and requires precise leveling and shimming. You might need to use shims of varying thicknesses to ensure the unit is perfectly plumb, level, and square. This meticulous approach prevents drafts and ensures the unit sits properly in the frame. Failing to shim adequately can lead to problems with closing mechanisms or cause future water damage. Remember, precision is key in this stage to avoid further complications.

After achieving the correct fit and placement, securing the unit is the next step. This involves using appropriate fasteners – such as nails or screws, depending on the door or window material. Remember to avoid over-tightening, which might damage the unit's frame. Ensure the fasteners are adequately placed and distribute the load evenly across the unit. Overlooking this step can weaken the unit's connection to the framing, creating vulnerabilities to weather and stress.

Finally, carefully apply weatherstripping and caulking around the unit to ensure airtightness and prevent water infiltration. This critical step greatly contributes to the overall energy efficiency and durability of your installation. Choose weatherstripping and caulking materials appropriate for the local climate and the materials used in the unit's construction. This often-overlooked step protects the interior of your structure from exposure to wind, rain, and temperature fluctuations.

Remember, proper installation isn't merely about getting the door or window in place; it's about creating a secure, weather-resistant, and functional seal. A well-installed door or window will significantly enhance the energy efficiency of a building, reducing heating and cooling costs. Careful attention to detail at each stage ensures optimal results. This process, while seemingly straightforward, demands precise execution. Inaccurate measurements, improper shimming, or insufficient sealing can all lead to considerable problems ranging from unsightly gaps to significant structural issues. Therefore, consistent attention to the details, precision in measurements, and a thorough understanding of the specific installation requirements for both the door or window and the framing are crucial for achieving a professional and lasting result. Always prioritize safety and use appropriate personal protective equipment (PPE) throughout the entire process.

6 Advanced Carpentry Techniques

Timber framing, a technique centuries old, demands precision and a deep understanding of wood's properties. Unlike modern methods relying on nails and screws, timber framing utilizes intricate joinery—mortise and tenon, dovetail, and others—to create strong, visually stunning structures. Mastering these joints requires patience, skill in using hand tools, and a keen eye for detail. The process begins with selecting the right timber, considering its species, grain, and potential for warping or cracking. Accurate measurements are paramount; even minor errors can compromise the entire structure's integrity. Each joint needs to be meticulously crafted, ensuring a perfect fit and maximizing the strength of the connection. The final assembly requires careful planning and execution, often involving a team of skilled craftsmen. The result is a structure that showcases the beauty of wood and stands as a testament to traditional building techniques.

Architectural detailing elevates a building beyond mere functionality, transforming it into a work of art. From intricate cornices and moldings to finely crafted window frames and door surrounds, these details add character and personality. Understanding classic architectural styles, whether it's the elegance of Georgian, the ornate charm of Victorian, or the clean lines of contemporary design, is crucial for creating harmonious and visually appealing details. This involves not just skillful carpentry but also an artistic sensibility, an ability to translate designs into three-dimensional form. Creating these details often requires a range of specialized tools, from carving chisels and molding planes to routers and shapers. Precision is key, as any imperfection will be immediately noticeable. Achieving a smooth, polished finish requires careful sanding, and potentially the application of various finishes to protect and enhance the wood's natural beauty or to achieve a specific aesthetic effect.

Restoration and repair of older structures presents unique challenges, requiring both technical skill and historical sensitivity. Often, you are dealing with aged, sometimes damaged, timber, and materials that may no longer be readily available. Thorough assessment of the existing structure is crucial, identifying areas needing repair or replacement while preserving the building's historical integrity. This may involve careful removal of damaged sections, matching existing joinery techniques, and sourcing replacement timbers that complement the originals. Understanding the history of building methods and the original materials used is essential. You may need to research historic techniques, such as using lime mortar instead of modern cement, or employing traditional methods of joinery. The goal isn't simply to fix the structure, but to restore it to its former glory, preserving its historical significance and architectural character for future generations. Patience and attention to detail are vital, as the work requires painstaking care and precision. Modern materials and techniques can sometimes be incorporated, but only when they are compatible with the historical integrity of the building and used thoughtfully, preserving its authenticity. This respect for the past ensures the longevity of a building's character and legacy.

The challenges of working with old structures extend beyond the mere repair of physical damage. Often, structural elements have settled or shifted over time, leading to potential instability. Understanding the forces acting on the building, and how to address these issues through careful reinforcement and stabilization, is a crucial part of any restoration project. This may involve introducing new structural members without compromising the original character, or utilizing traditional techniques like sistering beams to add strength while maintaining a cohesive visual appearance. The subtle art of blending old and new is key to a successful restoration, where modern techniques are used subtly, discreetly enhancing the integrity and longevity of the building without diminishing its historical charm. It demands not just carpentry skills, but also a keen understanding of building physics and historical preservation.

6.1 Timber Framing

Timber framing is a centuries-old craft, utilizing massive, hand-hewn timbers to create structures of immense strength and beauty. Unlike modern framing techniques employing standardized lumber and nails, timber framing relies on precise joinery, creating interlocked connections that form a strong and visually stunning skeleton. The process begins with careful selection of timbers, often large, straight trees, felled and milled to the exact dimensions needed for each component. This meticulous planning is key; a single miscalculation can compromise the entire structure.

Understanding the various joinery techniques used in timber framing is paramount. The mortise and tenon, while familiar from other carpentry, takes on a larger-than-life scale in timber framing. These joints, often reinforced with wooden pegs, must be executed with absolute precision, ensuring a perfect fit that distributes weight effectively. Other joints, such as the bridle joint and the saddle joint, offer variations suitable for different structural elements and aesthetic goals. These are not joints to be rushed; each demands patience and an unwavering focus on accuracy. The precision required is what makes timber framing so demanding, but also so rewarding.

Preparation of the timbers themselves is a crucial step, often involving hand-tool work. While power tools can assist in certain phases, the final shaping and fitting of the timbers typically involves hand planes, chisels, and saws. This creates both the structural integrity and the uniquely textured appearance for which timber framing is so admired. The careful shaping and smoothing of these large pieces is a testament to the skill and dedication of the craftsman. The craftsman's artistry is not just in the cutting, but in the final fitting and finishing.

Raising a timber frame is a spectacle, often involving a team working in concert. The weight of the timbers necessitates careful planning and execution. Using specialized techniques and tools like heavy-duty cranes or even traditional methods such as hand-heaving, each piece must be precisely placed and secured. This is where experience and teamwork are invaluable, coordinating the efforts of several people while working with potentially dangerous heavy materials. The coordination involved in the lifting of these beams is a spectacle in itself.

Once erected, the frame undergoes a further refinement. The joints are checked, any gaps are addressed, and the frame is prepared for the next stage of construction. Often, this involves the installation of infill, such as traditional wattle and daub or more modern insulation methods within the timber frame structure. This step bridges the gap between the traditional timber framing and modern building standards, creating a comfortable and energy-efficient structure. The integration of modern materials and techniques while preserving the integrity of the traditional approach is crucial.

Finally, the completion of the timber frame structure marks a significant milestone. The raw beauty of the exposed timbers offers a unique aesthetic. However, careful consideration should be given to protecting the timber from the elements. Appropriate finishes, such as natural oils or stains, can enhance the longevity of the structure while preserving its natural beauty, offering a balance between preservation and protection. The choice of finish must be carefully considered and selected to match the client's preferences and the requirements for the specific environment.

Timber framing, while labor-intensive and requiring specialized skills, offers unparalleled craftsmanship and enduring structural integrity. It's a testament to the connection between human ingenuity and the natural world, resulting in buildings that are not only beautiful but also built to withstand the test of time. The investment in time, skill, and materials is rewarded by a building that is both aesthetically pleasing and structurally superior, lasting for generations. Mastering this traditional technique allows for the creation of unique and sustainable structures. The feeling of accomplishment and the quality of the outcome are unparalleled. This method represents a timeless art, blending traditional methods with modern practicality. It provides a remarkable blend of sustainability and aesthetics for lasting beauty. By understanding and employing these techniques, you can contribute to a rich architectural legacy.

6.2 Architectural Details

Adding architectural details transforms a simple structure into a captivating masterpiece. These details, often overlooked in basic framing, are what truly elevate a building, injecting personality and character. They are not just aesthetic embellishments; carefully considered details enhance structural integrity and longevity. From intricate cornices to subtly crafted window casings, every element plays a crucial role.

Consider the cornice, a crowning feature that visually defines the roofline. A well-designed cornice not only adds aesthetic appeal but also provides essential protection to the building's structure by channeling rainwater away from the walls, preventing water damage and mold growth. Its projection, carefully calculated based on local climate conditions, determines its effectiveness and influence on the building's overall appearance. The construction, involving complex mitered cuts and possibly intricate molding, demands precise craftsmanship and an understanding of material properties. Choosing appropriate materials, like durable cedar or pressure-treated lumber, is vital for ensuring longevity. Remember to account for expansion and contraction of the materials due to seasonal temperature changes to prevent cracking or warping.

Next, let's focus on window and door casings. These seemingly simple elements contribute significantly to a building's visual appeal, offering opportunities for personalized expression. Casings frame the openings, drawing attention to them and adding a layer of visual depth. Their design can range from simple, clean lines for a minimalist look to ornate details incorporating moldings and carvings that reflect the architectural style of the home, providing a beautiful frame to doors and windows. Proper installation techniques are paramount, ensuring the casings are securely fastened to provide a weathertight seal and a professional finish that enhances structural integrity. Choosing the right material, such as durable hardwood or paintable composite, is a vital detail for longevity and the perfect finish.

Beyond casings, intricate moldings and trim offer endless opportunities for creative expression. These decorative elements add visual texture and depth, emphasizing architectural features and drawing the eye to particular points of interest. Crown molding, for example, adds a sophisticated touch, while baseboards protect the lower part of the walls and complement the overall design aesthetic. Careful planning and precise cuts are critical to a seamless, professional-looking installation. Proper measurements and the use of appropriate adhesives and fasteners will contribute to the durability of the installation and prevent the need for maintenance.

Balustrades and railings, if incorporated, present both aesthetic and safety considerations. Their design should complement the overall style of the building while ensuring sufficient strength and stability. Careful consideration of the materials used—from robust hardwoods to durable metals—is crucial. The spacing between balusters must adhere to building codes, ensuring the safety and security of occupants. Precise measurements are crucial for correct spacing and proper installation, preventing accidents and ensuring a long-lasting functional structure. The meticulous craftsmanship invested in these elements will showcase the skill of the carpenter.

Remember, architectural details are not merely decorative additions; they are integral parts of the building's design, contributing to its structural integrity, weather resistance, and overall aesthetic appeal. The careful selection of materials, precise measurements, and meticulous craftsmanship invested in these details will differentiate a simple structure from a truly remarkable building, showcasing the carpenter's skill and artistry. Thorough planning and attention to detail are essential for achieving a seamless and professional result that will last for generations. Mastering these techniques elevates your work from mere construction to the creation of lasting architectural works. Through careful planning and precise execution, you'll craft buildings that are not just functional but breathtakingly beautiful. This dedication to detail transforms your carpentry from a trade into an art form. The meticulous attention to finish work separates a good carpenter from a master craftsperson.

6.3 Restoration and Repair

Damaged wood often presents unique challenges. Unlike new lumber, you're working with material that's already experienced stress, possibly decay, and certainly has a history. Careful assessment is paramount; begin by identifying the extent of the damage. Is it superficial, impacting only the surface, or does it penetrate deeper into the structural core? Knowing this dictates the approach – a simple surface repair might suffice for minor scratches, whereas substantial rot necessitates complete replacement. Remember, safety always comes first. Wear appropriate personal protective equipment (PPE), including safety glasses, gloves, and a respirator, especially when dealing with potentially hazardous materials or deteriorated wood.

Understanding the type of wood is crucial. Different species react differently to damage and repair methods. Hardwoods, such as oak or mahogany, are more durable and often respond well to repair, but softwoods like pine are more prone to rot and require more attentive care. The age of the wood also plays a part; older wood might be more brittle and require gentler handling. Consider the existing finish; some finishes are more readily repaired than others, and some repair techniques are incompatible with certain finishes. Careful planning is always best for a successful outcome.

For minor surface damage, such as scratches or small gouges, sanding can often be enough. Start with a coarser grit sandpaper to remove the damaged area, then gradually work your way to finer grits to achieve a smooth finish. Once sanded, the area can be stained or finished to match the surrounding wood. For deeper damage, you may need to use wood filler. Choose a filler that closely matches the color and texture of the original wood. Apply the filler according to the manufacturer's instructions, ensuring it's level with the surrounding surface. After it dries, sand it smooth and finish as needed.

Repairing rot is a more complex endeavor. You'll need to remove all affected wood, back to sound, uncompromised material. This usually involves chiseling out the decayed areas. This requires careful observation; it's crucial to remove all compromised areas. Any remaining rot will undermine your repair and the problem will return. Once the rotten wood is removed, the void needs to be filled. Using epoxy resin is effective for this, ensuring a solid and durable repair that will last. Remember to let the epoxy cure fully before sanding and finishing.

Replacing entire structural members is sometimes unavoidable, particularly with significant damage or decay. This might involve removing and replacing a damaged beam, joist, or rafter. Always ensure the replacement member is of the same or superior quality and strength as the original. Accurate measurements and precise cuts are essential for a seamless fit. Pay attention to any joinery techniques; maintaining the structural integrity of the whole is critical. Proper fastening is paramount, using appropriate screws, nails, or bolts to secure the replacement in place.

When restoring old structures, you might find yourself dealing with antique hardware or unique architectural details. These require special care. Antique hinges or latches often need to be cleaned and oiled, sometimes requiring more intensive restoration. Consider using specialist cleaners for old metals. While preserving original features is always ideal, sometimes replacement is necessary. If you need to replace a piece, choose materials that match the originals as closely as possible, respecting the historical integrity of the structure.

Finally, after any restoration or repair work, it's essential to apply a protective finish. This prevents future damage and helps preserve the repaired area. Choosing the right finish depends on the wood type, location, and intended use. Oil-based finishes offer excellent protection, while water-based finishes are typically easier to apply and clean up. Ensure the finish is applied evenly to prevent any noticeable patchiness or uneven coloring. Following these guidelines will help you achieve successful and long-lasting restoration and repair work, breathing new life into old structures and wood components. Remember always to consult relevant building codes and regulations.

7 Commercial Construction Carpentry

Commercial construction carpentry demands a different scale of thinking than residential work. You're no longer building a single house; you're part of a larger team, constructing a complex structure with numerous interconnected parts. Understanding this shift in perspective is key to success. The sheer volume of materials and the intricate coordination required necessitates meticulous planning and a thorough understanding of building codes. This is where project management becomes paramount, a skill that needs practice and experience to truly master.

Effective project management starts with detailed blueprints and a comprehensive schedule. You need to know precisely what materials are needed, when they'll arrive, and how they'll be incorporated into the structure. This isn't simply a list; it's a dynamic plan, regularly updated to reflect changes and potential delays. This involves coordinating with subcontractors—electricians, plumbers, HVAC installers—ensuring everyone works smoothly and safely. Communication is the backbone of this process. Regular meetings, clear communication channels, and a proactive approach to problem-solving are vital for maintaining momentum and avoiding costly setbacks. Think of it as orchestrating a complex symphony where everyone plays their part perfectly.

Large-scale framing presents unique challenges. You're working with significantly larger dimensions and often more complex structural designs. The accuracy of your cuts and the precision of your assembly become even more critical to avoid structural weaknesses. Proper bracing and support become vital, especially during the initial stages of construction. The weight of the materials and the height of the structures demand careful consideration of safety protocols. Using heavy equipment such as cranes and forklifts necessitates additional training and a strict adherence to safety regulations. This isn't just about individual safety; it's about protecting your fellow workers.

Specialized techniques in commercial construction often involve advanced joinery methods, prefabricated components, and specialized tools. You may encounter engineered wood products, high-strength steel connections, and complex framing systems designed to meet specific engineering requirements. Familiarizing yourself with these techniques often necessitates additional training or mentorship from experienced professionals. These specialized skills set you apart and increase your earning potential, offering you access to higher-paying projects and more rewarding career opportunities. Learning never stops in this field.

Adhering to building codes and safety regulations is non-negotiable. Commercial construction is heavily regulated, ensuring the safety and structural integrity of the building. Ignoring these regulations can lead to costly fines, delays, and even legal action. Staying updated on the latest codes and regulations is crucial, necessitating continuous learning and attending relevant professional development courses. Familiarizing yourself with the local authorities and their requirements is also a very smart move. This means more than just reading the codes; it means understanding their practical implications in your work.

Beyond the technical aspects, commercial carpentry demands professionalism and leadership. You may supervise teams of carpenters, manage budgets, and deal with clients or contractors. Developing strong communication, negotiation, and problem-solving skills is essential. Managing expectations and meeting deadlines are hallmarks of a successful commercial carpenter. These aren't just skills; they are the foundations of a long and fulfilling career, transforming you from a skilled craftsman into a leader in the industry. Understanding this transition is fundamental. Your ability to guide and inspire others mirrors your own professional growth. It's a journey of continuous learning and improvement. The scale of the projects expands your horizons, and the challenges demand both technical skill and leadership capabilities. The rewards reflect this, providing both satisfaction and substantial financial growth.

7.1 Large-Scale Framing

Large-scale framing demands a different approach than residential projects. Forget the quick, intuitive methods you've used before; precision and planning become paramount. You're no longer dealing with a few dozen studs and joists; instead, you're working with hundreds, perhaps thousands, creating a skeleton of immense proportions. This necessitates meticulous pre-planning, involving detailed blueprints, accurate material lists, and a well-defined workflow. Think of it as orchestrating a symphony of wood, each piece playing its part in creating a harmonious whole.

Consider the sheer scale of materials. Efficient material handling is essential. Pre-cutting lumber on-site, though common in smaller projects, often becomes impractical. Fabricators specializing in large-scale projects can pre-cut and even pre-assemble components, drastically speeding up construction and minimizing waste. This proactive approach reduces on-site cutting time, enhancing safety and improving overall efficiency. You'll be surprised how much more smoothly the project flows with this level of organization.

Communication becomes critical. On a large project, you're part of a team; welders, electricians, plumbers, and other tradespeople rely on your work. Coordinating efforts requires clear communication channels and meticulous scheduling. Regular meetings, detailed drawings, and clearly marked components are essential for everyone to stay on track. A miscommunication here can lead to costly delays and setbacks. Think about the different elements of the project, from the foundation to the roof, and how they intersect and overlap.

Engineering principles take center stage. While basic framing relies on intuition and experience, large-scale construction demands precise calculations to ensure structural integrity. Load-bearing capacity, wind resistance, and seismic considerations become significant factors. Consult structural engineers; their expertise is vital for ensuring the long-term safety and stability of the structure. Your role isn't just about carpentry; you're contributing to the creation of a safe and durable building. Always prioritize safety; it's not an add-on, but an integral part of the process.

Consider the site itself. Access to materials and equipment may be more challenging, and transportation logistics require more thought. Site conditions could vary greatly affecting the project. Plan for potential issues; weather delays, equipment malfunctions, and logistical bottlenecks are all possibilities you must account for. These are often out of your control, but by being prepared you'll minimize the damage. Always maintain clear communication, especially when unexpected issues arise.

Beyond the technical aspects, leadership qualities are crucial. You will likely oversee a team, requiring strong organizational and communication skills. Delegating tasks effectively, monitoring progress, and maintaining a productive atmosphere are critical for a successful project. Remember, you are not just a carpenter but a leader. This involves inspiring the team and motivating them to achieve a common goal. Effective teamwork leads to a far superior result.

Finally, remember your own well-being. Large-scale projects demand long hours and intense concentration. Prioritizing rest, maintaining a healthy lifestyle, and fostering strong team dynamics are crucial for preventing burnout. Taking breaks, maintaining a positive attitude, and ensuring that all crew members are well-cared for are essential components of a project's success. This ensures both a positive team dynamic and a far safer work environment. In the end, a completed project is a testament not only to your carpentry skills but also your leadership and overall management of the large-scale endeavor.

7.2 Specialized Techniques

Working on large-scale commercial projects demands a different approach than residential construction. You'll encounter specialized materials and techniques unfamiliar in smaller jobs. Consider the use of prefabricated components; these aren't just for roofs—entire wall sections, even modular units, can be constructed off-site and assembled quickly on the job site, accelerating the project's timeline and minimizing on-site labor costs. This precision-engineered approach requires meticulous planning and coordination, highlighting the importance of detailed blueprints and seamless communication among the construction team.

Understanding and properly utilizing engineered wood products (EWP) is crucial. These materials, like laminated veneer lumber (LVL) and parallel strand lumber (PSL), offer superior strength and consistency compared to standard lumber, allowing for longer spans and thinner members, optimizing structural efficiency and often reducing material costs. However, working with EWPs demands familiarity with their specific handling and fastening requirements; improper handling can lead to structural weaknesses, so thorough knowledge is paramount.

Working with metal framing systems is another departure from traditional timber framing. Lightweight steel studs and tracks are frequently used in commercial buildings, particularly high-rises. They're lighter than wood, fire-resistant, and offer superior dimensional stability, but their installation requires specialized tools and an understanding of their unique connection methods, often involving self-drilling screws and specialized fasteners. Remember safety precautions, as sharp edges and potential pinch points are common.

Formwork construction, crucial for concrete structures, constitutes a major aspect of commercial carpentry. Building robust, accurate forms for walls, columns, and beams necessitates proficiency in various techniques like panel construction, bracing, and shoring. Understanding the properties of different concrete mixes and their impact on formwork design is vital. Moreover, you must consider the considerable pressure exerted by wet concrete on forms, ensuring stability throughout the curing process, to prevent collapse and costly rework.

In commercial settings, you frequently encounter curtain wall systems, large exterior wall assemblies incorporating glass and metal panels. Installation demands precision and attention to detail, emphasizing watertightness and structural integrity. Often, these systems are prefabricated off-site, emphasizing effective site management and careful assembly by the carpentry team, requiring careful handling and a thorough understanding of the system's unique connections and tolerances.

Then there's the matter of specialized finishes. Commercial projects might involve complex millwork, including intricate paneling, custom cabinetry, and detailed trim. This demands advanced woodworking skills, often incorporating specialized machinery and finishing techniques. Precision is vital; imperfections stand out in a larger building, and these are the details that distinguish quality craftsmanship. Such projects often involve close collaboration with designers and architects, making effective communication essential.

Beyond the technical aspects, successful commercial carpentry hinges on understanding building codes and safety regulations. This is critical. Compliance is non-negotiable; it necessitates staying abreast of the specific codes applicable to your region, including those concerning fire safety, accessibility, and structural requirements. Safety protocols must be scrupulously followed, as commercial sites generally involve larger teams and more significant risks. Your awareness and diligence contribute directly to a secure and efficient work environment for everyone on the job site. This requires constant vigilance and a deep understanding of regulations.

Finally, understanding project management principles within the commercial construction context is invaluable. Larger projects involve complex scheduling, material ordering, and coordination with other trades. The ability to manage your time effectively, forecast material needs accurately, and collaborate smoothly with other teams contributes significantly to project success and timely completion. This goes beyond just carpentry; it's about managing a piece of the entire project, which is essential for your own success and that of the entire project. Effective communication and problem-solving skills are absolutely crucial for navigating the complexities of a large-scale commercial project.

7.3 Building Codes

Understanding building codes is paramount for any carpenter, regardless of whether you're building a small shed or a large commercial complex. These aren't just arbitrary rules; they're safety regulations designed to protect the occupants and the structure itself. Ignorance of building codes can lead to significant legal issues, project delays, and, most importantly, potential harm. Therefore, familiarity with your local codes is not optional, it's a necessity. You must actively seek out this information and study it diligently. Your local building department is an excellent starting point; they'll have all the relevant documentation readily available, often online.

Beyond the basics of safety, building codes delve into specific requirements for various aspects of construction. For example, the spacing of studs in a wall isn't just a matter of preference; codes dictate minimum spacing based on the wall's height and intended load-bearing capacity. Similarly, the type and size of lumber you use must meet minimum strength criteria to ensure structural integrity and withstand anticipated stresses. Don't assume shortcuts are permissible. In fact, many codes specifically prohibit them, leaving you open to penalties if caught. Always refer to the specific code requirements before making any decisions that deviate from standard practices. Remember, seemingly minor details can have a significant cumulative impact on the overall safety and stability of the structure.

Electrical work is heavily regulated in building codes. Every aspect, from wire gauge selection to outlet placement, is precisely defined and must be followed to the letter. Incorrect wiring can result in serious hazards, such as fires or electrocution. While you might feel qualified to perform simple electrical tasks, it is almost always recommended to engage a licensed electrician for this type of work, ensuring all electrical requirements are correctly met and documented. This not only protects you from liability, but it also ensures compliance with codes, safeguarding everyone involved. The potential consequences of electrical code violations are substantial.

Plumbing codes are similarly stringent. They dictate the type of pipes to be used, the proper installation techniques, and the required slope for drainage to ensure a functioning and safe plumbing system. Leaking pipes can damage the structure and lead to health hazards. Similar to electrical work, unless you have undergone the necessary training and licensing, it's advisable to let a licensed plumber handle your plumbing work. Their expertise ensures compliance with codes, eliminating potential problems before they arise. This proactive approach protects your building's structural integrity and the health of its future occupants.

Beyond the mechanical, aspects of building design are equally subject to building codes. The precise requirements vary wildly from location to location. For instance, local building codes will determine the minimum requirements for insulation, helping to ensure energy efficiency and maintain a comfortable interior climate. These codes are frequently updated, so staying informed about the latest regulations is crucial. You might find regularly checking the updates online to be a necessary practice, or even attending workshops or seminars on current building regulations can significantly enhance your awareness. Compliance with these aspects safeguards both the building's performance and its inhabitants' well-being.

Finally, always maintain accurate records. This is crucial, not only for personal reference but also for compliance checks. Keep detailed records of all materials used, inspections passed, and any modifications made during the construction process. These documents are essential in case of any disputes or inspections. Organized documentation, including permits and approvals, provides a clear audit trail, demonstrating your compliance with building codes. The value of comprehensive documentation should never be underestimated. In the event of an issue, well-maintained records can greatly assist in resolving disputes and protecting your interests. Remember that meticulous record-keeping is a crucial aspect of responsible and compliant construction practices. It also reflects your professionalism as a builder.

7.4 Project Management

Effective project management in carpentry, especially for commercial projects, hinges on meticulous planning and execution. Ignoring this can lead to cost overruns, schedule delays, and compromised quality – potentially jeopardizing the entire project. A solid plan acts as your roadmap, guiding you through every stage, from initial design to final completion. You'll need to have a keen eye for detail and an ability to adapt; flexibility is vital in the face of unexpected challenges.

Your first step is developing a comprehensive project plan. This isn't just a simple to-do list; it's a detailed breakdown of every task involved, outlining the sequence of operations, resource allocation, and timeframes. Consider using a Gantt chart or similar visual tool to represent the project schedule clearly. Detailed scheduling aids visualization of tasks' dependencies; it becomes clear what needs to be completed before the next phase can start. This helps anticipate potential bottlenecks before they occur. Make sure you have considered factors like material delivery times and potential weather delays – those unforeseen problems can dramatically impact the timeline.

Next, you must create a budget. This involves estimating the cost of all materials, labor, and equipment. It is critical to build in a contingency fund to absorb unforeseen expenses. Thorough research is necessary to acquire accurate material pricing. Factor in labor costs based on the skills required and the number of workers needed for different phases. Don't forget equipment rental fees if you're outsourcing some tasks. Accurate budgeting will also help when it comes time to seek financing if needed. It also helps manage expectations with clients, ensuring transparency from the outset.

Effective communication is paramount. Keep the client, subcontractors, and your team informed at each stage. Regular meetings, detailed progress reports, and transparent updates prevent miscommunication and maintain a smooth workflow. Clearly defining roles and responsibilities at the beginning helps prevent duplication of efforts and ensures accountability. A shared online project management tool can streamline communication, helping to centralize project information and facilitate easier collaboration. Regular communication also builds trust, which is essential for a successful outcome.

Throughout the project, you must monitor progress closely and make necessary adjustments. This involves tracking actual costs against the budget, measuring productivity, and identifying potential problems. Regular site inspections, progress reviews with the client and team members, and careful tracking of labor hours are crucial for staying on schedule and within budget. Address issues promptly to prevent minor problems from escalating into major setbacks. By constantly monitoring the project, you can proactively adapt to challenges, ensuring that the overall timeline remains achievable.

Finally, remember the importance of health and safety. Enforce safety protocols consistently throughout the project to minimize risks and ensure the well-being of your team. Regular safety checks, training, and provision of adequate personal protective equipment (PPE) are non-negotiable for reducing accidents. This commitment also helps demonstrate professionalism, reinforcing your reputation as a reliable and responsible contractor. This attention to safety not only safeguards your crew, but also contributes significantly to project efficiency by minimizing disruptions caused by accidents. Regular safety inspections should become ingrained within your project management methodology.

Effective project management isn't solely about following a checklist; it's about cultivating a proactive and adaptable approach. By understanding the intricacies of planning, budgeting, communication, monitoring, and safety, you can transform complex projects into successful undertakings. Remember, a successful project is not merely one that's completed on time and within budget; it's also one completed with safety as the utmost priority and with the satisfaction of both your team and the client. The ability to seamlessly manage all aspects of the project distinguishes the competent carpenter from the truly exceptional one. It is in this meticulous approach that the true essence of mastery lies. Mastering these project management skills will elevate your work and help you achieve your career goals. Careful planning prevents costly mistakes and ensures smooth construction, leading to client satisfaction and a strong reputation. Ultimately, good project management equates to quality workmanship and business success. Don't underestimate the power of preparedness and proactive problem-solving in this crucial aspect of carpentry.

7.5 Safety Regulations

Working on commercial construction projects necessitates a heightened awareness of safety regulations, exceeding the basic precautions taken on smaller residential jobs. These regulations are not mere suggestions; they are legally mandated requirements designed to protect workers and the public. Ignoring them can lead to hefty fines, project shutdowns, and, most importantly, serious injury or death. Understanding and adhering to these regulations is paramount, a responsibility that every carpenter, from apprentice to foreman, must take seriously.

Your primary responsibility involves personal safety. This isn't just about wearing a hard hat—although that's a vital starting point—it's a holistic approach that encompasses every aspect of your work. Properly functioning safety harnesses are critical when working at heights, ensuring you're securely anchored to prevent falls. Eye protection is non-negotiable, protecting you from flying debris and harmful particles generated by power tools. Hearing protection is just as crucial, mitigating long-term hearing damage from the constant noise associated with construction equipment. Finally, and perhaps most commonly overlooked, appropriate hand protection is vital. Gloves prevent cuts, abrasions, and exposure to harmful substances.

Beyond personal protection, you must ensure the safety of others on the job site. This involves understanding and following site-specific safety plans and adhering to all posted warnings. Regular site inspections are critical; you must be vigilant in identifying and reporting any hazards, no matter how seemingly insignificant. This proactive approach prevents accidents before they occur. Furthermore, always use the correct tools for the job, making sure the tools are in good working order. Improper tool usage is a leading cause of accidents, and any faulty equipment requires immediate attention. Ignoring these issues can have catastrophic consequences, affecting not only your safety but the safety of your co-workers.

Maintaining a clean and organized workspace is a significant safety precaution. Cluttered areas create trip hazards and obstruct movement, increasing the likelihood of falls and accidents. Tools should be stored properly when not in use, preventing accidents and ensuring efficiency. Keeping walkways clear is vital, particularly in busy construction environments where materials and workers move constantly. Clear communication also plays a role; always alert others of your actions, particularly if working in close proximity to heavy equipment or elevated work areas. Simple precautions can greatly improve site safety.

Beyond personal and workplace safety, you'll need familiarity with local, state, and federal regulations, which vary widely and can be quite extensive. It's your responsibility to stay abreast of the most current laws and guidelines. This often involves regularly reviewing updated codes and attending safety training sessions provided by your employer or industry organizations. These sessions provide essential updates on new safety procedures, changes to regulations, and best practices. Don't overlook these; they're vital for your continued safety and the safety of others. Staying informed is a continuous process, demanding commitment and proactive engagement.

Furthermore, understanding and applying specific safety protocols for the use of machinery is crucial. Heavy equipment, like cranes and forklifts, requires certified operators who understand both the equipment and the necessary safety procedures. You should never operate such equipment unless you're fully trained and certified to do so. Similar rules apply to power tools; knowing the safe operating procedures and understanding their limitations will prevent numerous hazards. It's not just enough to know how to use them; you must use them safely.

Finally, remember, safety isn't just a set of rules; it's a mindset. It's a commitment to working diligently and carefully in every task, however mundane it may seem. Always maintain a heightened sense of awareness, ensuring that safety considerations remain at the forefront of your mind at all times. You're responsible not only for your own well-being but also for the well-being of your colleagues and anyone else who might be present on the job site. Embrace safety as a core principle of your profession.

www.ingramcontent.com/pod-product-compliance
Lightning Source LLC
LaVergne TN
LVHW022021200926
848481LV00007B/2482
9798348565725